Chapters

Chapters   **1**

Chapter 1: Introduction   **4**

    Explanation of the concept of reality   4

    Why understanding reality is important   5

Chapter 2: The Nature of Reality   **7**

    Different theories of reality   8

    The debate over whether reality is objective or subjective   9

    The role of perception in shaping our understanding of
reality   10

Chapter 3: The Limits of Perception   **12**

    The limitations of our senses   12

    How our brain interprets sensory information   14

    The concept of the "reality tunnel"   15

Chapter 4: The Illusion of Time   **17**

    The nature of time   17

    The debate over whether time is real or an illusion   18

    The implications of the idea that time is an illusion   19

Chapter 5: The Illusion of Space   **21**

    The nature of space   21

    The debate over whether space is real or an illusion   23

    The implications of the idea that space is an illusion   24

Chapter 6: The Role of Consciousness   **25**

    The relationship between consciousness and reality   25

    The debate over whether consciousness creates reality or
merely perceives it   27

    The implications of the idea that consciousness creates
reality   28

Chapter 7: The Multiverse   **30**

The concept of the multiverse   31

The different types of multiverse theories   32

The implications of the idea that there are multiple realities 33

Chapter 8: The Simulation Hypothesis   **35**

The concept of the simulation hypothesis   35

The evidence for and against the simulation hypothesis   36

The implications of the idea that reality is a simulation   37

Chapter 9: The Nature of Matter   **39**

The debate over whether matter is real or an illusion   39

The different theories of the nature of matter   40

The implications of the idea that matter is an illusion   41

Chapter 10: The Quantum World   **43**

The basics of quantum mechanics   43

The implications of quantum mechanics for our understanding of reality   44

The different interpretations of quantum mechanics   45

Chapter 11: The Relationship between Mind and Body   **47**

The mind-body problem   47

The different theories of the relationship between mind and body   48

The implications of the idea that mind and body are separate entities   49

Chapter 12: The Concept of Free Will   **51**

The debate over whether free will exists   51

The different theories of free will   52

The implications of the idea that free will does or does not exist   53

Chapter 13: The Meaning of Life   **55**

The different theories of the meaning of life   55

The implications of the idea that life has no inherent
meaning   57
The role of individual perspective in creating meaning   57
Chapter 14: The Future of Reality   **59**
The potential implications of future technology on our
understanding of reality   60
The different scenarios for the future of reality   61
The role of human consciousness in shaping the future of
reality   62
Chapter 15: Conclusion   **64**
Recap of the book's main points   64
Final thoughts on the nature of reality and its implications 65
Call to action for readers to continue exploring the truth
about reality.   66

Chapter 1: Introduction

Reality is a fundamental concept that has fascinated philosophers, scientists, and ordinary people for centuries. It is the foundation upon which we build our understanding of the world and ourselves. Despite its importance, there is still much that remains unknown and debated about the nature of reality. This book sets out to explore the truth about reality and to shed light on the many questions that surround it.

In this introductory chapter, we will provide an overview of what reality is and why understanding it is essential. We will also give a brief preview of the topics that will be covered in the following chapters. From the limits of perception to the nature of matter and the role of consciousness, this book will explore some of the most significant and debated concepts related to reality.

Ultimately, the goal of this book is to challenge readers to think critically about their own understanding of reality and to encourage them to continue exploring this fascinating and complex topic. Whether you are a philosopher, scientist, or simply curious about the world around you, this book will provide you with insights and perspectives that will deepen your understanding of reality and its implications for our lives.

Explanation of the concept of reality

At its most basic level, reality refers to everything that exists, whether it is physical, mental, or abstract. It encompasses everything we experience through our senses, including objects, people, emotions, and ideas. Reality is the foundation upon which we build our understanding of the world, and it shapes our beliefs, values, and actions.

The concept of reality has been the subject of philosophical inquiry for centuries. Philosophers have debated the nature of reality and how we can know anything about it. One of the most fundamental questions is whether reality is objective or subjective. Objective reality refers to a reality that exists independently of our perceptions, while subjective reality is based on our individual experiences and perceptions.

The concept of reality is also closely related to the idea of truth. Truth refers to a state of affairs that accurately reflects the way things are in reality. In other words, if something is true, it corresponds to reality. However, determining what is true can be challenging, as our perceptions can be biased, incomplete, or distorted.

In addition to the debate over the nature of reality and truth, there are also questions about the scope of reality. For example, some theories suggest that reality is limited to what we can directly experience through our senses, while others propose the existence of unseen dimensions or realms beyond our perception.

Overall, the concept of reality is complex and multifaceted, and there is no single, definitive answer to what it is. However, exploring the many different perspectives on reality can deepen our understanding of the world and ourselves, and provide insights into some of the most profound questions about our existence.

Why understanding reality is important

Understanding reality is crucial because it forms the basis for our beliefs, values, and actions. It allows us to make sense of the world and navigate through it successfully. Furthermore, an accurate understanding of reality can help us solve problems, make informed decisions, and improve our quality of life.

For example, our understanding of reality influences how we approach scientific research and development. Scientists rely on an accurate understanding of the nature of reality to develop new technologies, test hypotheses, and make discoveries. Without a solid understanding of reality, scientific progress would be hindered, and we would be limited in our ability to solve problems and improve our lives.

Similarly, an understanding of reality is essential in many other areas of life, including ethics, politics, and spirituality. Our beliefs about reality influence our moral and ethical values, and guide our decision-making processes. In politics, our understanding of reality shapes our views on important issues such as social justice, human rights, and the environment. In spirituality, our understanding of reality can guide our beliefs about the afterlife, the existence of a higher power, and the purpose of our existence.

Finally, an accurate understanding of reality can help us cope with the challenges and uncertainties of life. By understanding the nature of reality, we can develop a sense of perspective and meaning that can help us navigate through difficult times and find purpose in our lives.

In conclusion, understanding reality is essential for our personal and collective well-being. It forms the foundation upon which we build our beliefs, values, and actions, and guides our decisions in many areas of life. By exploring the nature of reality, we can deepen our understanding of the world and ourselves, and find new ways to solve problems, make informed decisions, and improve our lives.

Chapter 2: The Nature of Reality

At the heart of our exploration of reality is the question of what it actually is. In this chapter, we will delve into the nature of reality itself and explore some of the different theories that have been proposed to explain it.

One of the most fundamental debates is whether reality is objective or subjective. Objective reality refers to a reality that exists independently of our perceptions, while subjective reality is based on our individual experiences and perceptions. Some argue that reality is entirely subjective and that there is no objective reality, while others believe that objective reality is the foundation upon which all other realities are built.

We will also examine the question of whether reality is deterministic or indeterministic. A deterministic reality is one in which all events are predetermined and can be predicted with complete accuracy. In contrast, an indeterministic reality suggests that there is an element of randomness and unpredictability in the universe.

Furthermore, we will explore some of the different metaphysical theories that have been proposed to explain the nature of reality, including idealism, materialism, and dualism. Each of these theories offers a unique perspective on the nature of reality and the relationship between mind and matter.

Ultimately, the question of the nature of reality is a complex and multi-faceted one. By examining the different perspectives and theories, we can gain a deeper understanding of this fundamental concept and the implications it has for our lives and the world around us.

Different theories of reality

Throughout history, many different theories have been proposed to explain the nature of reality. These theories range from metaphysical philosophies to scientific theories and offer a variety of perspectives on what reality is and how it works. In this section, we will explore some of the most significant theories of reality.

1. Idealism: Idealism is the theory that all of reality is ultimately mental or spiritual in nature. According to idealism, the physical world we experience is an illusion, and reality is ultimately composed of consciousness or ideas. Some philosophers have argued that reality is a product of our minds, while others have suggested that reality exists independently of our perception but is ultimately composed of consciousness or spiritual essence.

2. Materialism: Materialism is the theory that reality is entirely composed of matter and energy. According to this view, everything in the universe can be reduced to its fundamental physical components. Materialism is the basis for much of modern scientific thinking, and it has been a driving force in the development of many technological advances.

3. Dualism: Dualism is the theory that reality is composed of two distinct substances, mind and matter. According to this view, the mind and the physical body are separate entities that interact with each other but are fundamentally different in nature. Some philosophers have suggested that the mind is a non-physical substance that exists independently of the body, while others have argued that the mind and body are both physical substances but that they are distinct from each other.

4. Holism: Holism is the theory that reality is fundamentally interconnected and that everything in the universe is part of a larger whole. According to holism, the universe is more than the sum of its parts, and it cannot be understood by breaking it down into its component parts. Holism is often associated with spiritual and mystical traditions, but it has also been incorporated into scientific and ecological thinking.
5. Constructivism: Constructivism is the theory that reality is a social construction and that our perception of the world is shaped by our social and cultural context. According to constructivism, our understanding of reality is not objective but is instead a product of the language, symbols, and norms of our culture.

In conclusion, there are many different theories of reality, each offering a unique perspective on what it is and how it works. Exploring these theories can deepen our understanding of the world and ourselves and provide insights into some of the most profound questions about our existence.

The debate over whether reality is objective or subjective

One of the most fundamental debates in the study of reality is whether it is objective or subjective. Objective reality refers to a reality that exists independently of our perceptions, while subjective reality is based on our individual experiences and perceptions.

Those who argue for an objective reality believe that there is a fundamental truth that exists beyond our subjective experiences. They argue that reality exists independently of our perceptions and that our understanding of it is limited by our own biases and limitations. In this view, reality is considered to be objective, whether we are aware of it or not.

On the other hand, those who argue for a subjective reality believe that reality is constructed by our individual perceptions and experiences. They argue that there is no objective reality outside of our subjective experiences and that our understanding of reality is always filtered through our own perceptions and biases. According to this view, what is real for one person may not be real for another, and reality is ultimately subjective.

The debate over whether reality is objective or subjective has far-reaching implications for many different fields, including science, philosophy, and spirituality. For example, in science, the question of whether reality is objective or subjective has important implications for the nature of scientific inquiry and the relationship between the observer and the observed. In philosophy, the debate over the nature of reality has been ongoing for centuries, with some arguing for an objective reality that exists independently of our perceptions, while others argue that reality is fundamentally subjective.

Ultimately, the question of whether reality is objective or subjective is a complex one, and there is no easy answer. However, exploring the different perspectives and theories can provide valuable insights into the nature of reality and how it impacts our understanding of the world around us.

The role of perception in shaping our understanding of reality

Our perception plays a critical role in shaping our understanding of reality. How we perceive the world around us affects how we interpret and make sense of our experiences. Perception is not just a passive process of receiving sensory information, but an active process of interpreting and making sense of that information.

Our perception is shaped by many factors, including our past experiences, cultural background, and personal biases. For example, our experiences and cultural background can shape our beliefs about what is real and what is not. Our personal biases can also influence how we perceive information and can lead us to interpret things in a way that confirms our preexisting beliefs.

The role of perception in shaping our understanding of reality can also be seen in the phenomenon of optical illusions. Optical illusions occur when our brains interpret sensory information in a way that does not match reality. These illusions demonstrate the active nature of perception and highlight how our perception can be influenced by factors beyond our control.

The importance of perception in shaping our understanding of reality has been recognized by many philosophers and scientists. Some argue that our perception is the only way we can access reality, and therefore, our understanding of reality is fundamentally limited by the limitations of our senses. Others suggest that our perception is influenced by our cultural and personal biases, and that our understanding of reality is always limited by these biases.

In conclusion, perception plays a critical role in shaping our understanding of reality. Our perception is not just a passive process of receiving sensory information, but an active process of interpreting and making sense of that information. The importance of perception in shaping our understanding of reality has important implications for fields ranging from psychology to philosophy, and it is a topic of ongoing debate and exploration.

Chapter 3: The Limits of Perception

As we've seen in the previous chapter, perception plays a critical role in shaping our understanding of reality. However, our perception has its limits, and there are aspects of reality that we simply cannot perceive. These limitations can be due to physical constraints, such as the limitations of our senses, or to cognitive constraints, such as the limitations of our attention and memory.

Understanding the limits of perception is important because it helps us to recognize the biases and limitations that are inherent in our understanding of reality. It also helps us to appreciate the vastness and complexity of the world around us, and to recognize the importance of interdisciplinary approaches to understanding reality.

In this chapter, we will explore the limits of perception and the ways in which they shape our understanding of reality. We will examine the physical constraints that limit our perception, such as the limitations of our senses and the scale of the universe. We will also look at the cognitive constraints that limit our perception, such as the limitations of our attention and memory, and the ways in which our perception can be influenced by our expectations and beliefs.

By understanding the limits of perception, we can gain a deeper appreciation for the complexity and mystery of the world around us, and recognize the importance of humility in our understanding of reality.

The limitations of our senses

Our senses are our primary means of perceiving the world around us. However, our senses have limitations that can prevent us from perceiving certain aspects of reality. For example, we can only see a limited range of wavelengths of light, which means that we cannot see certain colors or objects that emit radiation outside of that range.

Similarly, our sense of hearing is limited to a certain range of frequencies, which means that we cannot hear sounds that are too high or too low in pitch. Our sense of smell is also limited, and we cannot detect certain odors that are too faint or outside of our range of sensitivity.

Our sense of touch is limited in its ability to detect fine details or changes in temperature, pressure, or texture. Our sense of taste is limited in its ability to distinguish between different flavors, and our ability to perceive taste can also be influenced by our cultural background and personal preferences.

These limitations of our senses can have important implications for our understanding of reality. For example, the limitations of our visual perception can prevent us from perceiving certain objects or phenomena, such as infrared radiation or subatomic particles. Similarly, our limited ability to detect changes in temperature or pressure can make it difficult for us to perceive certain environmental changes or detect certain dangers.

Despite these limitations, our senses are still a valuable tool for perceiving the world around us. We can use technology to extend the range of our senses, such as telescopes to see objects that are too far away or microscopes to see objects that are too small to be seen with the naked eye.

In conclusion, the limitations of our senses are an important factor in shaping our understanding of reality. Our senses provide us with a valuable tool for perceiving the world around us, but we must also recognize their limitations and the ways in which they can bias our understanding of reality. Understanding the limitations of our senses is an important step in recognizing the complexity and mystery of the world around us.

How our brain interprets sensory information

Our senses provide us with a stream of information about the world around us. However, the information provided by our senses is not always accurate or complete. Our brain must interpret and make sense of this information to create a coherent perception of reality.

The process of interpreting sensory information begins in the sensory receptors, which detect and transduce physical stimuli into neural signals that can be processed by the brain. These neural signals are then transmitted to the brain's primary sensory areas, where they are integrated and processed to form a representation of the sensory input.

However, this representation is not a simple reflection of the physical world. Our brain uses a variety of processes to interpret and make sense of the sensory information it receives. These processes can be influenced by our expectations, past experiences, and attentional focus.

For example, our brain uses a process called top-down processing, where it uses our expectations and past experiences to interpret sensory information. This can lead to perceptual biases, such as seeing familiar patterns or shapes in ambiguous stimuli.

Our brain also uses selective attention to focus on certain aspects of the sensory input and ignore others. This can lead to the filtering out of certain information that may be important for a complete understanding of the situation.

The interpretation of sensory information is also influenced by the context in which it occurs. For example, the same sensory input can be interpreted differently depending on the surrounding environment or the task at hand.

In conclusion, our brain plays a critical role in interpreting sensory information to create a perception of reality. The interpretation process is influenced by a variety of factors, including our expectations, past experiences, and attentional focus. Understanding how our brain interprets sensory information can help us to recognize the biases and limitations inherent in our perception of reality.

The concept of the "reality tunnel"

The concept of the "reality tunnel" refers to the idea that our perception of reality is shaped by our individual experiences, beliefs, and biases. This concept was first introduced by Timothy Leary, a psychologist and writer who explored the effects of psychedelic drugs on consciousness.

According to Leary, our perception of reality is shaped by a "tunnel" of our own experiences and beliefs, through which we view and interpret the world around us. This "tunnel" can limit our perception of reality and prevent us from seeing alternative perspectives or possibilities.

The concept of the "reality tunnel" has been further developed by other writers and thinkers, who have applied it to a variety of fields, including psychology, philosophy, and spirituality. The idea is that our individual reality tunnels can shape our beliefs, values, and behavior, and can limit our ability to see alternative viewpoints or possibilities.

For example, if someone grows up in a particular cultural or religious context, their reality tunnel may be shaped by the beliefs and values of that context. This can lead to a limited understanding of other cultures or beliefs, and can prevent the individual from seeing alternative perspectives.

Similarly, our reality tunnel can be influenced by our biases, such as confirmation bias or availability bias. These biases can limit our perception of reality by leading us to selectively focus on information that confirms our preexisting beliefs or that is readily available to us.

In conclusion, the concept of the "reality tunnel" highlights the ways in which our perception of reality is shaped by our individual experiences, beliefs, and biases. By recognizing the limitations of our reality tunnel, we can broaden our perspective and gain a deeper understanding of the complex and multifaceted nature of reality.

Chapter 4: The Illusion of Time

Time is a fundamental aspect of our experience of reality. We perceive time as a linear progression, with events unfolding in a sequential order. However, recent scientific research has challenged this common-sense view of time and has revealed that our perception of time may be an illusion.

The concept of time has been a subject of philosophical inquiry for centuries, with many different theories about the nature of time and its relationship to reality. Some philosophers have argued that time is a fundamental aspect of reality, while others have claimed that time is an illusion created by our minds.

In this chapter, we will explore the concept of time and the evidence that suggests that our perception of time may be an illusion. We will examine the various theories about the nature of time, including the block universe theory, the presentism theory, and the growing block universe theory. We will also discuss the implications of the illusion of time for our understanding of reality and our place within it.

The nature of time

The nature of time has been a subject of philosophical inquiry for centuries. Philosophers have debated whether time is a fundamental aspect of reality or simply an illusion created by our minds. There are various theories about the nature of time, including presentism, eternalism, and growing block universe theory.

Presentism is the view that only the present moment exists, and the past and future are not real. According to this theory, time is constantly moving forward, and the present is the only moment that truly exists. All events and experiences that occur in the past or future are only ideas in our minds and do not have any actual reality.

Eternalism, on the other hand, is the view that past, present, and future all exist simultaneously. This theory suggests that all moments in time exist in a timeless "block universe" and are equally real. According to this view, our perception of time as a linear progression is an illusion.

The growing block universe theory suggests that only the past and present exist, and the future is continually being created. This theory suggests that as time moves forward, the universe grows and changes, creating new moments of time.

Despite these different theories, the nature of time remains a mystery. Scientific research has challenged our common-sense view of time and has revealed that our perception of time may be an illusion. Our experience of time may be influenced by factors such as our perception of the speed of events, our memory, and our attention.

In conclusion, the nature of time remains a subject of debate and inquiry. While different theories offer different perspectives on the nature of time, the idea that our perception of time may be an illusion challenges our understanding of reality and raises questions about the nature of existence itself.

The debate over whether time is real or an illusion

The debate over whether time is real or an illusion has been a subject of philosophical inquiry for centuries. Some philosophers argue that time is a fundamental aspect of reality, while others claim that time is merely an illusion created by our minds.

Those who argue that time is real point to the fact that we experience the passage of time in our daily lives. We observe changes in the world around us, and we remember events that have occurred in the past. These experiences suggest that time is a fundamental aspect of reality.

On the other hand, those who argue that time is an illusion point to the fact that our experience of time is subjective and can be influenced by various factors. For example, time seems to slow down when we are engaged in an activity that requires our attention, and it seems to speed up when we are bored or disengaged. Additionally, our perception of time can be influenced by our memory, emotions, and physical state.

Scientific research has also challenged the idea that time is a fundamental aspect of reality. Einstein's theory of relativity suggests that time is relative and can vary depending on an observer's position and velocity. This theory suggests that time may not be an objective feature of reality but may be an illusion created by our minds.

Ultimately, the debate over whether time is real or an illusion may be a matter of perspective. Our experience of time is subjective and can be influenced by various factors, but that does not necessarily mean that time is not real. Regardless of whether time is ultimately determined to be real or an illusion, the concept of time remains a fundamental aspect of our experience of reality.

The implications of the idea that time is an illusion

The idea that time is an illusion challenges our understanding of reality and raises important philosophical and scientific questions. If time is an illusion, what implications does this have for our understanding of the world and our place in it?

One implication of the idea that time is an illusion is that it challenges our common-sense view of causality. Causality suggests that events in the past cause events in the future, but if time is an illusion, this raises questions about how causality can be understood. If all moments in time exist simultaneously, then it is not clear how one moment can cause another.

Another implication of the idea that time is an illusion is that it challenges our understanding of personal identity. Our experience of time is intimately tied to our sense of self, and if time is an illusion, then our sense of self may be called into question. If all moments in time exist simultaneously, then it is not clear how we can be said to exist over time as a continuous and distinct entity.

The idea that time is an illusion also has implications for our understanding of the nature of reality. If time is an illusion, then what else about our experience of reality may be illusory? Our perception of the world around us is intimately tied to our experience of time, and if time is an illusion, then it is possible that our perception of the world may also be influenced by illusions.

Finally, the idea that time is an illusion raises important scientific questions about the nature of the universe. If time is an illusion, then what is the ultimate nature of reality? What is the underlying structure of the universe, and how can we come to understand it?

In conclusion, the idea that time is an illusion challenges our understanding of reality and raises important philosophical and scientific questions. While the implications of this idea may be unsettling, they also offer new avenues for inquiry and a deeper understanding of the nature of existence.

Chapter 5: The Illusion of Space

Just as the concept of time has been the subject of philosophical inquiry for centuries, so too has the concept of space. We tend to think of space as a fundamental aspect of reality, a feature of the world that exists independently of our perception of it. However, recent scientific research has challenged this assumption and suggested that space may be an illusion created by our minds.

The idea that space is an illusion challenges our understanding of reality and raises important questions about the nature of the world around us. If space is an illusion, then what is the underlying structure of the universe, and how can we come to understand it? What implications does this have for our understanding of physics and the laws that govern the universe?

In this chapter, we will explore the idea that space may be an illusion and the implications of this idea for our understanding of reality. We will examine scientific research that supports this idea, as well as philosophical arguments that have been made in favor of it.

Ultimately, the idea that space may be an illusion challenges us to rethink our understanding of the world and to question our assumptions about the nature of reality. By exploring this idea and its implications, we can deepen our understanding of the universe and our place within it.

The nature of space

Space is commonly thought of as a fundamental aspect of reality, a feature of the world that exists independently of our perception of it. We often think of space as a container for objects, a framework within which physical events occur. However, recent scientific research has suggested that our understanding of space may be an illusion created by our minds.

According to traditional physics, space is three-dimensional and Euclidean, meaning that it follows the rules of Euclidean geometry. Objects within this space are said to have a specific location and can move through space along a defined trajectory. However, in the early 20th century, Einstein's theory of relativity challenged this understanding of space by suggesting that space and time are intertwined and that the structure of space-time can be influenced by the presence of massive objects.

More recently, scientists have explored the idea that space may be an emergent property of the universe, rather than a fundamental aspect of it. According to this idea, space is not a pre-existing structure, but rather arises from the interactions of fundamental particles and fields. This notion is consistent with some theories of quantum gravity, which suggest that space and time are emergent properties that arise from a deeper, underlying structure.

The idea that space may be an illusion created by our minds has also been explored by philosophers and mystics throughout history. Some argue that our perception of space is an artifact of our senses and that, in reality, there is no such thing as a "spatial" world. Others suggest that space may be a construct of our consciousness, a way of organizing our experience of the world.

In conclusion, the nature of space is a complex and contested topic that challenges our assumptions about the nature of reality. While traditional physics has conceived of space as a fundamental aspect of the universe, recent scientific research and philosophical inquiry suggest that space may be an emergent property or an illusion created by our minds. By exploring these different perspectives on space, we can deepen our understanding of the nature of the universe and the limitations of our perception.

The debate over whether space is real or an illusion

The question of whether space is real or an illusion has been the subject of debate among scientists and philosophers for centuries. While traditional physics has treated space as a fundamental aspect of reality, recent scientific research and philosophical inquiry have challenged this assumption and suggested that space may be an illusion created by our minds.

One argument in favor of the idea that space is an illusion is based on the concept of non-locality in quantum mechanics. According to this concept, particles can be connected in a way that transcends space and time, such that actions taken on one particle can instantaneously affect the state of another particle, regardless of their physical distance from each other. This suggests that the concept of physical distance in space may be an illusion, and that the universe may be fundamentally non-local.

Another argument in favor of the idea that space is an illusion is based on the concept of holographic principle. According to this principle, the information contained in a three-dimensional space can be encoded in a two-dimensional surface. This suggests that our perception of three-dimensional space may be an illusion, and that the underlying reality may be two-dimensional.

On the other hand, many scientists and philosophers argue that space is a fundamental aspect of reality, and that the concept of an illusionary space is a result of our limited perception. They argue that our perception of space is necessary for us to navigate and interact with the world around us, and that without it, we would not be able to function in the physical world.

Ultimately, the debate over whether space is real or an illusion challenges us to rethink our understanding of reality and our assumptions about the nature of the universe. While scientific research and philosophical inquiry have provided compelling arguments on both sides of the debate, the question of whether space is an illusion or a fundamental aspect of reality remains unresolved.

The implications of the idea that space is an illusion

The idea that space is an illusion has significant implications for our understanding of reality and our place in the universe. If space is an illusion, it would mean that the physical world we experience is not as objective as we once thought, and that our perception of space is limited by our minds.

One implication of this idea is that our perception of distance and size may not accurately reflect the underlying reality. Objects that appear far away may not be as distant as we think, and objects that appear large may not be as big as we perceive them to be. This challenges our common-sense understanding of the physical world and suggests that our perception of space may be a construction of our minds.

Another implication of the idea that space is an illusion is that our understanding of cause and effect may need to be reevaluated. The concept of non-locality in quantum mechanics challenges our understanding of how actions lead to consequences in the physical world, and suggests that the universe may operate according to different principles than we previously thought.

Furthermore, the idea that space is an illusion challenges us to consider the nature of consciousness and its relationship to the physical world. If space is a construction of our minds, then our perception of reality may be more subjective than objective, and our experience of the world may be intimately tied to our consciousness.

Overall, the idea that space is an illusion challenges us to question our assumptions about the nature of reality and to consider new ways of understanding the universe. While the implications of this idea may be unsettling, they also provide an opportunity for us to expand our understanding of the world and to explore new avenues of scientific and philosophical inquiry.

Chapter 6: The Role of Consciousness

The nature of consciousness has been a topic of debate among philosophers and scientists for centuries. While consciousness is a fundamental aspect of our experience, it remains a mysterious and elusive phenomenon that is not yet fully understood. In recent years, however, there has been a growing interest in the role of consciousness in shaping our understanding of reality.

This chapter will explore the role of consciousness in our perception of reality and the ways in which it influences our understanding of the world around us. We will examine various theories of consciousness, from the traditional dualistic view that separates the mind and body to the more recent panpsychist view that suggests that consciousness is a fundamental aspect of the universe.

We will also explore the ways in which our consciousness shapes our perception of reality, from the limitations of our senses to the role of attention and intention in shaping our experience. Additionally, we will examine the relationship between consciousness and the physical world, including the possibility that consciousness plays a fundamental role in the creation and maintenance of the universe.

Overall, this chapter aims to provide a comprehensive overview of the role of consciousness in shaping our understanding of reality. By exploring the various theories and implications of consciousness, we can gain a deeper appreciation for the complexity and mystery of the human experience.

The relationship between consciousness and reality

The relationship between consciousness and reality is a topic of great interest to both scientists and philosophers. While there is no consensus on the nature of consciousness or its role in the physical world, there are a number of theories that attempt to explain this relationship.

One theory is the idea of panpsychism, which suggests that consciousness is a fundamental aspect of the universe and exists at all levels of reality, from subatomic particles to the human mind. This theory suggests that consciousness plays a role in the creation and maintenance of the physical world, and that our perception of reality is intimately tied to our consciousness.

Another theory is the idea of idealism, which suggests that the physical world is a product of our minds and that consciousness is the ultimate reality. According to this view, the world exists only as it is perceived by conscious beings, and there is no objective reality independent of consciousness.

Other theories suggest that consciousness arises from the physical processes in the brain, and that there is no direct relationship between consciousness and the external world. These theories suggest that our perception of reality is the result of the processing of sensory information by the brain, and that consciousness is an emergent property of this process.

Regardless of the specific theory, there is a growing body of evidence to suggest that our consciousness plays a fundamental role in shaping our perception of reality. Studies have shown that our expectations and beliefs can influence our sensory experiences, and that our attention and intention can influence the outcome of physical experiments.

Overall, the relationship between consciousness and reality remains a topic of ongoing investigation and debate. While there is no definitive answer to the question of how consciousness and reality are related, exploring this topic can deepen our understanding of the nature of consciousness and the complexity of the physical world.

The debate over whether consciousness creates reality or merely perceives it

One of the most contentious debates in the study of consciousness and reality is whether consciousness actively creates reality or merely perceives it. This debate has important implications for our understanding of the nature of consciousness and the physical world.

Proponents of the idea that consciousness creates reality suggest that our perception of the world is not simply a passive reflection of an objective reality, but that consciousness actively shapes and creates the world around us. This view is often associated with the concept of the "observer effect" in quantum physics, which suggests that the act of observing a system can actually change its behavior.

Those who hold this view suggest that consciousness is a fundamental aspect of the universe, and that our thoughts and intentions have a direct impact on the physical world. This idea has gained popularity in recent years, with the rise of books and films like "The Secret" and "What the Bleep Do We Know?!" which suggest that our thoughts and beliefs can shape our reality.

On the other hand, critics of this view suggest that consciousness is merely a passive observer of reality, and that our perception of the world is largely determined by the physical processes in the brain. They argue that while our thoughts and beliefs may influence our behavior, they have little or no impact on the external world.

This view is often associated with the materialist paradigm in science, which suggests that the physical world is the ultimate reality and that consciousness is simply a byproduct of physical processes. According to this view, consciousness is an epiphenomenon of the brain, and has no causal power over the physical world.

Despite the ongoing debate over whether consciousness creates reality or merely perceives it, there is growing evidence to suggest that our beliefs and expectations can have a profound impact on our perception of the world. Studies have shown that our mental states can influence everything from our physical health to the outcome of social interactions.

Ultimately, the debate over whether consciousness creates reality or merely perceives it is a complex and multifaceted issue that is likely to continue to be a subject of ongoing investigation and debate in the years to come.

The implications of the idea that consciousness creates reality

The idea that consciousness actively creates reality has far-reaching implications for our understanding of the nature of reality and our place in the universe. If consciousness is indeed a fundamental aspect of the universe, then our thoughts and intentions have the potential to shape and influence the physical world in profound ways.

One of the most significant implications of this idea is that we may have more control over our lives than we previously thought. If our thoughts and beliefs have the power to shape our reality, then we have the ability to consciously direct our lives in ways that we may not have realized before.

This idea has given rise to a number of practices and techniques designed to help individuals harness the power of their thoughts and beliefs to manifest their desires and achieve their goals. These practices, which are often referred to as "manifestation" or "law of attraction" techniques, involve visualizing desired outcomes and focusing one's thoughts and intentions on them in order to bring them into reality.

The idea that consciousness creates reality also challenges our traditional notions of causality and determinism. If consciousness has the power to influence the physical world, then the idea of cause and effect becomes more complex. It suggests that our thoughts and intentions may play a role in determining the outcome of events, rather than being entirely determined by preceding physical causes.

This idea also has important implications for our understanding of the mind-body connection and the relationship between mental and physical health. If our thoughts and beliefs have the power to influence the physical world, then they may also have a direct impact on our physical well-being. This idea has led to the development of new approaches to health and healing that emphasize the importance of addressing mental and emotional factors in addition to physical symptoms.

Overall, the idea that consciousness creates reality challenges our traditional understanding of the relationship between mind and matter, and opens up new possibilities for understanding and exploring the nature of reality. While the debate over this idea is likely to continue, it has already had a profound impact on our understanding of the power of human consciousness and its potential to shape the world around us.

Chapter 7: The Multiverse

For centuries, humans have pondered the possibility of parallel universes, alternate realities, and multiple versions of ourselves existing simultaneously. While this idea was once relegated to the realm of science fiction, recent developments in physics have given rise to the theory of the multiverse – a concept that proposes the existence of multiple, coexisting universes, each with their own unique physical laws and properties.

The theory of the multiverse is a controversial and complex subject, with many different interpretations and variations. Some scientists argue that the existence of a multiverse is a logical consequence of certain theories in physics, such as string theory or inflationary cosmology. Others suggest that the multiverse is a purely speculative concept that lacks empirical evidence and cannot be tested.

Despite the debate, the idea of the multiverse has captured the imagination of scientists and laypeople alike, with many grappling with the implications of such a concept. If the multiverse exists, it could mean that there are countless versions of ourselves living out different life paths and making different choices. It could also mean that there are other worlds where the laws of physics are vastly different, and where life as we know it may not even be possible.

In this chapter, we will explore the theory of the multiverse in depth, examining the different variations and interpretations of the concept, as well as the evidence that supports or refutes it. We will also delve into the philosophical and existential implications of the multiverse, including its potential impact on our understanding of free will, determinism, and the nature of reality itself.

The concept of the multiverse

The concept of the multiverse is based on the idea that there may be more than one universe, each with its own set of physical laws and properties. This idea has been explored by many theoretical physicists, and it has become an increasingly popular topic in the field of cosmology.

The basic idea behind the multiverse is that our universe is just one of many possible universes that exist. Each of these universes would have different properties, such as different values for the fundamental constants of nature, or different numbers of dimensions. Some versions of the multiverse theory suggest that there could be an infinite number of these universes, while others propose a finite number.

One version of the multiverse theory is based on the concept of "eternal inflation." This theory suggests that the universe is expanding at an accelerating rate, and that this expansion will continue indefinitely. As a result, new universes could be created in this inflationary process, with each new universe having its own unique properties.

Another version of the multiverse theory is based on the idea of the "many-worlds" interpretation of quantum mechanics. According to this theory, every time a quantum measurement is made, the universe splits into multiple universes, each corresponding to a different possible outcome of the measurement. This would mean that every possible outcome of any quantum measurement is realized in a separate universe.

While the concept of the multiverse is intriguing, it is important to note that it is still a theoretical idea that has not been proven or directly observed. Some scientists have criticized the idea of the multiverse, arguing that it is untestable and therefore unscientific.

Despite these criticisms, the concept of the multiverse continues to capture the imagination of many people. If it is true, the multiverse could fundamentally change our understanding of the universe and our place within it. It could mean that there are infinite versions of ourselves living out different life paths, and it could provide a possible explanation for some of the mysteries of the universe, such as the so-called "fine-tuning" of the fundamental constants of nature.

The different types of multiverse theories

There are several different types of multiverse theories that have been proposed by scientists and philosophers. Here are some of the most well-known:

1. The Level I multiverse: This is the simplest type of multiverse theory, which is based on the idea of an infinite universe. According to this theory, if the universe is infinite, then there must be an infinite number of regions within it that are identical to our observable universe.

2. The Level II multiverse: This theory is based on the concept of inflation, which suggests that the universe underwent a period of rapid expansion in its early history. According to this theory, the inflationary process may have created multiple "bubble" universes, each with its own set of physical laws and properties.

3. The Level III multiverse: This theory is based on the idea of the "many-worlds" interpretation of quantum mechanics. According to this theory, every time a quantum measurement is made, the universe splits into multiple universes, each corresponding to a different possible outcome of the measurement.

4. The Level IV multiverse: This theory is based on the idea of the landscape of string theory, which suggests that there may be an enormous number of different possible universes that could exist, each with its own unique set of physical laws and properties.
5. The Mathematical multiverse: This theory is based on the idea that the universe is fundamentally mathematical in nature, and that there are an infinite number of mathematical structures that exist independently of any physical universe.

Each of these multiverse theories has its own unique implications and challenges, and scientists continue to explore and debate the feasibility and implications of each theory. While the concept of the multiverse remains a controversial and unproven idea, it is a fascinating and thought-provoking area of study that challenges our understanding of reality and our place within it.

The implications of the idea that there are multiple realities

The idea that there are multiple realities, or a multiverse, has profound implications for our understanding of the universe and our place within it. Here are some of the key implications of this idea:

1. The existence of multiple realities would suggest that our observable universe is just one small part of a much larger and more complex reality. This would challenge our intuition and force us to rethink our understanding of the nature of existence and the universe as a whole.
2. The idea of multiple realities would also have important implications for the concept of free will. If there are an infinite number of possible universes, each corresponding to a different possible outcome of any given situation, then it could be argued that free will is an illusion, and that everything that happens is predetermined by the laws of physics and the initial conditions of the universe.

3. The concept of the multiverse would also have implications for the anthropic principle, which suggests that the universe seems finely tuned to support life because we happen to exist in one of the rare universes that allows for our existence. If there are an infinite number of universes, each with its own set of physical laws and properties, then it is possible that there are many other universes that also support life, and that the apparent fine-tuning of our universe is simply a result of selection bias.

4. Finally, the concept of multiple realities would challenge our understanding of the nature of reality itself. If there are an infinite number of possible universes, each with its own unique set of physical laws and properties, then it could be argued that reality is not fixed or objective, but rather subjective and dependent on the observer.

Overall, the idea of multiple realities is a fascinating and provocative area of study that raises important questions about the nature of the universe and our place within it. While the concept remains controversial and unproven, it challenges us to think deeply about the fundamental nature of existence and the universe as a whole.

Chapter 8: The Simulation Hypothesis

The idea that we might be living in a simulated reality, similar to a computer program, has captured the imagination of many scientists, philosophers, and science fiction writers. This idea, known as the simulation hypothesis, raises fascinating questions about the nature of reality and our place within it.

In this chapter, we will explore the simulation hypothesis in depth, examining its origins, its implications, and the evidence for and against it. We will consider what it would mean for us to be living in a simulated reality, and how we might be able to test or prove this hypothesis.

The simulation hypothesis has important implications for a wide range of fields, including physics, philosophy, and psychology. It challenges our understanding of what is real and what is not, and raises profound questions about the nature of existence and the universe as a whole.

Join us as we delve into the fascinating world of the simulation hypothesis, exploring its implications and its potential to reshape our understanding of reality.

The concept of the simulation hypothesis

The simulation hypothesis suggests that our reality may be a computer-generated simulation, much like the virtual worlds created in video games or other simulations. The idea is that our reality is not the fundamental reality, but rather a simulated reality created by an advanced civilization, possibly in the future.

This concept has been popularized by philosophers and scientists, such as Nick Bostrom, who argue that if a civilization reaches a certain level of technological advancement, it would have the capability to create a simulated reality that is indistinguishable from the real world. They argue that if such a civilization exists, it is more likely that we are living in a simulation rather than in the "real" world.

The simulation hypothesis raises profound questions about the nature of reality and our place within it. If we are living in a simulation, then what is the purpose of our existence? Is there a way to break out of the simulation and discover the true reality? And if so, how would we do it?

While the simulation hypothesis is often considered to be a far-fetched idea, it has gained some support from scientists and thinkers who argue that it is a possibility that should be taken seriously. Some scientists have even proposed methods for testing the hypothesis, such as looking for glitches in the simulation or searching for evidence of a "programmer" behind the scenes.

Regardless of whether the simulation hypothesis is ultimately proven to be true or false, it challenges our assumptions about reality and encourages us to think deeply about the nature of existence and our place within the universe.

The evidence for and against the simulation hypothesis

While the simulation hypothesis is an intriguing concept, it is difficult to test or prove definitively. Nevertheless, proponents and skeptics have put forth evidence and arguments for and against the idea.

One argument in favor of the simulation hypothesis is that as technology continues to advance, we may eventually reach a point where we are able to create realistic simulations ourselves. If that is the case, then it stands to reason that a more advanced civilization could have already created a simulation of our reality.

Another argument in favor of the simulation hypothesis is that there are some phenomena that seem to defy explanation in our current understanding of the laws of physics. For example, some scientists have proposed that the apparent "quantum entanglement" of particles across vast distances could be explained by a simulated reality, where information is transmitted instantaneously across the simulation.

However, there are also arguments against the simulation hypothesis. One is that the complexity and detail of our reality seems to be far beyond what we could create with current technology, let alone what we can imagine. Additionally, some skeptics argue that even if it is possible to create a simulated reality, there would be no reason for an advanced civilization to do so.

Furthermore, some scientists have pointed to evidence that suggests our reality is not a simulation. For example, the discovery of the cosmic microwave background radiation, which is believed to be the residual heat left over from the Big Bang, seems to be difficult to explain within a simulated reality framework.

Overall, while there is no concrete evidence for or against the simulation hypothesis, it remains a thought-provoking and fascinating idea that challenges our understanding of reality and the universe.

The implications of the idea that reality is a simulation

If reality is indeed a simulation, then it would have profound implications for our understanding of ourselves and the universe. Here are some of the possible implications:

1. The nature of reality: If reality is a simulation, then what we perceive as "real" is actually a virtual construct. This would mean that the universe as we know it is not a physical, objective reality but a simulation created by an unknown entity.
2. The concept of free will: If reality is a simulation, then the idea of free will becomes questionable. If the simulation is predetermined or programmed, then our actions and choices may be predetermined as well.
3. The existence of a higher power: The concept of a "creator" or "higher power" takes on new meaning if reality is a simulation. The creators of the simulation could be seen as gods or as a highly advanced civilization.

4. The possibility of multiple simulations: If our reality is a simulation, then it's possible that there are multiple simulations, each with its own set of rules and parameters. This could mean that our reality is just one of many possible simulations.
5. The potential for manipulating reality: If reality is a simulation, then it may be possible to manipulate it in ways that we cannot currently conceive. This could have implications for fields such as medicine, technology, and even entertainment.

Overall, the idea that reality is a simulation challenges our fundamental understanding of the universe and our place within it. While it may seem far-fetched, it remains an intriguing concept that encourages us to question our assumptions and explore new possibilities.

Chapter 9: The Nature of Matter

The concept of matter has been a fundamental part of our understanding of reality since ancient times. Matter is the physical substance that makes up everything we can touch, see, and interact with in the world around us. From the smallest subatomic particles to the largest celestial bodies, matter is what gives the universe its substance and structure.

However, as our understanding of the universe has grown, so has our understanding of the nature of matter. The study of particle physics has revealed a world of subatomic particles that behave in strange and unexpected ways, challenging our traditional ideas of what matter is and how it behaves.

In this chapter, we will explore the nature of matter and the latest theories in particle physics. We will delve into the building blocks of matter, the fundamental forces that govern their behavior, and the mysteries that still remain unsolved. Through this exploration, we will gain a deeper understanding of the nature of reality and our place within it.

The debate over whether matter is real or an illusion

The debate over whether matter is real or an illusion has been a topic of philosophical inquiry for centuries. On one hand, materialists argue that matter is the fundamental building block of the universe and that it exists independently of human consciousness. They maintain that matter has an objective reality and that it can be studied and understood through scientific observation and experimentation.

On the other hand, some philosophers and scientists argue that matter may be an illusion created by our perceptions and consciousness. They point to the strange behavior of subatomic particles and the mysteries of quantum mechanics as evidence that our understanding of matter may be incomplete or even incorrect.

One popular interpretation of quantum mechanics is the idea that matter does not exist in a definite state until it is observed or measured. This has led some to propose that matter may be a product of our observations or consciousness, rather than an objective reality that exists independently of us.

However, while the debate over the nature of matter continues, it is important to note that regardless of its ultimate nature, matter still plays a vital role in our understanding and experience of the world around us. Whether it is an objective reality or a product of our perceptions, matter still shapes our interactions with the universe and provides the physical substance of our lives.

The different theories of the nature of matter

There are several different theories about the nature of matter that have been proposed throughout history. Here are a few of the most significant:

1. Atomism: The ancient Greek philosopher Democritus proposed the idea that all matter is made up of indivisible particles called atoms. This theory was later refined by other philosophers and scientists, including John Dalton in the 19th century.
2. Materialism: Materialism is the idea that matter is the fundamental substance of the universe, and that it exists independently of consciousness. This theory has been influential in scientific and philosophical thought for centuries.
3. Idealism: Idealism is the opposite of materialism, and suggests that consciousness is the fundamental substance of the universe, and that matter is an illusion created by our perceptions.
4. Dualism: Dualism is the idea that the universe is made up of two distinct types of substance: matter and consciousness. This theory has been debated by philosophers and scientists for centuries.

5. Quantum mechanics: The field of quantum mechanics has led to new theories about the nature of matter, including the idea that particles do not exist in a definite state until they are observed or measured.

While these theories may seem to be in conflict with one another, they all offer different perspectives on the nature of matter and its role in the universe. Each theory provides a unique way of understanding and explaining the physical world, and scientists and philosophers continue to explore these ideas in order to gain a deeper understanding of the universe we inhabit.

The implications of the idea that matter is an illusion

If matter is an illusion, then it would mean that the physical world we see and experience is not real. This concept is often associated with Eastern philosophical traditions, such as Hinduism and Buddhism, which suggest that the material world is maya, or an illusion.

If matter is an illusion, then the implications are significant. It would mean that our bodies, the objects around us, and even the laws of physics are not truly real. This can be a challenging concept to grasp, as it seems to contradict our direct sensory experience of the world.

However, some argue that this idea is supported by scientific theories such as quantum mechanics, which suggests that particles do not have a fixed position until they are observed. This implies that the act of observation itself affects the reality of the observed object, suggesting that our perception plays a crucial role in shaping reality.

If matter is an illusion, it could also mean that our consciousness plays a more fundamental role in creating reality than previously thought. This idea is consistent with certain spiritual and metaphysical beliefs that posit that consciousness is the underlying fabric of the universe.

Overall, the idea that matter is an illusion challenges our traditional understanding of the physical world and invites us to consider the possibility that our perception and consciousness are key factors in shaping reality.

Chapter 10: The Quantum World

The world of quantum mechanics is one of the most fascinating and enigmatic areas of science. It challenges our everyday notions of reality and introduces a strange and bewildering set of rules that govern the behavior of matter and energy on a microscopic scale. In this chapter, we will explore the weird and wonderful world of quantum mechanics and its implications for our understanding of reality.

We will delve into the mysteries of quantum entanglement, wave-particle duality, and the uncertainty principle, and examine the fascinating experiments that have revealed the bizarre behavior of particles at the quantum level. We will also explore the implications of quantum mechanics for our understanding of the nature of reality, including the role of the observer and the idea of a multiverse.

By the end of this chapter, you will have gained a deeper appreciation for the mind-bending concepts of quantum mechanics and the impact they have on our understanding of the world around us. So, let's take a journey into the strange and fascinating world of the quantum.

The basics of quantum mechanics

Quantum mechanics is a branch of physics that studies the behavior of matter and energy on a very small scale, such as the behavior of atoms and subatomic particles. At this scale, classical mechanics, which governs the behavior of objects we can see and touch, no longer applies.

In quantum mechanics, particles do not have definite positions or velocities until they are measured or observed. Instead, they exist in a state of superposition, meaning they can be in multiple states at once. This concept is known as wave-particle duality, which suggests that particles can also exhibit wave-like behavior.

Additionally, quantum mechanics introduces the idea of uncertainty, which states that certain properties of a particle, such as its position and momentum, cannot be measured simultaneously with perfect accuracy. This concept is described by Heisenberg's uncertainty principle.

Quantum mechanics also includes the idea of entanglement, where particles can become linked in such a way that their states are correlated, even when separated by great distances.

Overall, quantum mechanics presents a fascinating and complex view of the world, and its implications have yet to be fully understood.

The implications of quantum mechanics for our understanding of reality

The implications of quantum mechanics for our understanding of reality are profound and far-reaching. One of the most significant implications is that the classical picture of a deterministic universe, where objects exist in a definite state at all times, is incorrect. In the quantum world, particles can exist in multiple states simultaneously, known as superposition, until they are observed or measured.

This concept of superposition challenges our understanding of the nature of reality, as it suggests that the act of observation can fundamentally alter the state of an object. Additionally, the concept of entanglement, where particles can become linked in such a way that the state of one particle affects the state of the other, suggests that our understanding of locality and causality may need to be re-evaluated.

Quantum mechanics also poses challenges to our understanding of the nature of matter and the nature of space and time. The concept of wave-particle duality suggests that particles can exhibit both wave-like and particle-like behavior, blurring the lines between what we traditionally think of as matter and energy. The uncertainty principle suggests that it is impossible to simultaneously measure both the position and momentum of a particle with absolute accuracy, which challenges our understanding of the precise nature of space and time.

Overall, quantum mechanics challenges our understanding of reality in ways that are still being explored and understood by scientists and philosophers alike. The implications of this field of study are profound and may have far-reaching implications for our understanding of the nature of reality itself.

The different interpretations of quantum mechanics

There are several interpretations of quantum mechanics that attempt to explain the peculiar behaviors of subatomic particles. One of the earliest and most well-known interpretations is the Copenhagen interpretation, which holds that the act of measurement collapses the wave function and determines the outcome of an observation. This interpretation emphasizes the role of the observer in the measurement process.

Another interpretation is the Many-Worlds interpretation, which posits that every possible outcome of a quantum measurement actually occurs in a separate, parallel universe. This theory suggests that the universe is constantly branching into countless parallel realities, each corresponding to a different outcome of a quantum event.

A third interpretation is the Pilot-Wave theory, also known as the De Broglie-Bohm theory, which suggests that particles have definite positions and velocities at all times, but are guided by a pilot wave that determines their behavior. This theory eliminates the role of randomness in quantum mechanics and provides a deterministic explanation for quantum phenomena.

The implications of these interpretations are profound and far-reaching. They challenge our traditional understanding of cause and effect, the nature of reality, and the role of the observer in scientific inquiry. They also raise questions about the limits of human knowledge and our ability to comprehend the fundamental nature of the universe.

Chapter 11: The Relationship between Mind and Body

The mind-body problem is one of the most debated and fascinating topics in philosophy and neuroscience. It involves the question of how the mind and the body are related to each other, and whether they are separate entities or two aspects of the same thing. This chapter will explore the different theories about the mind-body problem and their implications for our understanding of reality.

The mind-body problem has been debated for centuries, with philosophers and scientists proposing various theories to explain the relationship between the mind and body. Some theories suggest that the mind and body are separate entities that interact with each other, while others propose that the mind and body are two aspects of the same thing.

The implications of the mind-body problem go beyond philosophy and neuroscience, as they have significant implications for our understanding of the nature of reality, the human experience, and even the existence of consciousness itself. This chapter will delve into these implications and explore the different theories that have been proposed to solve the mind-body problem.

The mind-body problem

The mind-body problem is one of the most enduring and controversial issues in philosophy and cognitive science. It concerns the nature of the relationship between the mind and the body, and asks whether the mind and body are separate entities or whether they are closely intertwined. On one hand, there is the view that the mind and body are distinct entities that interact with each other, while on the other hand, there is the view that the mind and body are one and the same.

The mind-body problem raises a number of important questions, such as: How does the mind interact with the body? Does the mind have any causal influence on the body? Can the mind exist independently of the body? These questions have important implications for our understanding of consciousness, free will, and the nature of reality.

Over the centuries, various philosophical and scientific perspectives have been proposed to tackle the mind-body problem. Some have argued for a dualist perspective, where the mind and body are seen as separate entities that interact with each other. Others have proposed a monist perspective, where the mind and body are seen as different aspects of the same entity. Still, others have suggested more nuanced positions, such as idealism, where the mind is seen as the primary reality and the body is a product of the mind.

In this chapter, we will explore the different perspectives on the mind-body problem and their implications for our understanding of reality. We will examine the arguments and evidence for each perspective and consider the strengths and weaknesses of each approach. Ultimately, the mind-body problem is one of the most fascinating and challenging issues in philosophy and science, and a deeper understanding of it can lead to profound insights about the nature of existence.

The different theories of the relationship between mind and body

There have been various theories proposed throughout history regarding the relationship between the mind and the body. One of the most fundamental debates is whether the mind and body are separate entities or whether they are interconnected and inseparable.

The dualist perspective, proposed by René Descartes in the seventeenth century, posits that the mind and body are two separate substances. According to this view, the mind is a non-physical substance that can exist independently of the body. Dualists argue that mental events cannot be reduced to physical events in the brain and that consciousness is a unique phenomenon that cannot be explained by physical processes.

On the other hand, monist theories propose that the mind and body are different aspects of the same thing. For example, materialism asserts that everything in existence, including the mind and consciousness, can be explained in terms of physical matter and processes. In contrast, idealism argues that everything is fundamentally mental, and that physical matter is an illusion created by the mind.

Another influential theory in the mind-body debate is functionalism, which suggests that mental states are defined by their functional roles or relationships with other mental states and with behavior. In this view, it doesn't matter what type of substance or material the mind is made of, as long as it performs the necessary functions.

Overall, the relationship between the mind and body remains a topic of active debate and research, with many different theories and perspectives to consider.

The implications of the idea that mind and body are separate entities

The idea that mind and body are separate entities has been a subject of philosophical and scientific debate for centuries. One of the main implications of this idea is the question of how the mind and body interact with each other. If the mind and body are separate entities, then how do they influence each other? Theories of dualism, which suggest that the mind and body are distinct entities, propose various mechanisms of interaction, such as through a non-physical substance like Descartes' concept of the pineal gland.

Additionally, the idea of mind-body dualism has implications for fields such as medicine and psychology. If the mind and body are separate, then treating physical symptoms may not address underlying psychological or emotional issues, and vice versa. This has led to the development of holistic approaches to health that take into account both physical and mental well-being.

Another implication of the mind-body problem is the concept of free will. If the mind and body are separate, then how does the mind's will influence the body's actions? This raises questions about the nature of decision-making and the extent to which our choices are predetermined by physical processes in the brain.

Overall, the concept of mind-body dualism has significant implications for our understanding of human nature and the relationship between our physical and mental experiences. While there is no consensus on the nature of this relationship, exploring the various theories and implications can provide valuable insights into the complexities of consciousness and the human experience.

Chapter 12: The Concept of Free Will

The idea of free will is one of the oldest and most controversial concepts in philosophy. It is the idea that humans have the ability to make choices that are not predetermined by any outside force, but are rather the result of our own personal agency. The concept of free will has implications for our understanding of morality, responsibility, and even the nature of reality itself.

In this chapter, we will explore the different theories of free will, the evidence for and against it, and the implications of the idea that we do, or do not, possess free will. We will examine the debate between determinism and indeterminism, the role of neuroscience and genetics in shaping our choices, and the implications of a world without free will. Ultimately, we will attempt to answer the question of whether or not we truly possess free will, and what that means for our understanding of ourselves and the world around us.

The debate over whether free will exists

The question of whether we have free will is one of the most fundamental and longstanding debates in philosophy. At its core, the debate centers around the extent to which our actions are determined by factors beyond our control, such as genetics, environment, and past experiences. Some argue that free will is an illusion, and that all our choices are ultimately predetermined by these external factors. Others maintain that we have the capacity to make choices that are genuinely free, and that our decisions are not predetermined by any outside forces.

One of the main arguments against the existence of free will is the idea of determinism. This theory holds that every event, including human actions, is caused by preceding events and natural laws. According to determinism, our decisions are ultimately predetermined by factors beyond our control, such as our genetic makeup and the environment we were raised in. Therefore, some argue, we do not have true free will.

On the other hand, advocates of free will argue that our decisions are not determined solely by external factors, and that we have the ability to make choices that are not entirely predetermined. They point to the fact that we often face situations where there is no clear external pressure or constraint on our decisions, and that in these instances, we are free to choose one course of action over another. Some also argue that the subjective experience of making choices is evidence of our capacity for free will, as it would not make sense to experience the sensation of making choices if we did not have the ability to do so.

The debate over the existence of free will has important implications for many areas of life, including ethics, law, and personal responsibility. If free will is an illusion, then it may be difficult to hold individuals fully responsible for their actions, as their behavior is ultimately predetermined by factors beyond their control. Conversely, if we do have free will, then we may be held accountable for our actions in a more robust way, as our choices are seen as a reflection of our own agency and decision-making capacity.

The different theories of free will

The concept of free will has been debated for centuries, and there are a variety of theories about its nature. One of the most prominent theories is libertarianism, which asserts that individuals have the ability to make choices that are not determined by external or internal factors. In other words, libertarianism suggests that humans have the ability to act freely and independently.

On the other hand, determinism is a theory that asserts that every event, including human actions, is predetermined by previous causes. This means that humans may not have the ability to make choices that are truly independent, as their actions are predetermined by factors beyond their control.

Compatibilism is another theory that seeks to reconcile free will with determinism. According to this theory, humans may still have the ability to make choices that are free, even if their actions are ultimately determined by prior causes.

Finally, some philosophers and scientists argue that free will is an illusion. They suggest that all human actions are ultimately determined by factors outside of our control, such as genetics and environmental factors. This idea challenges the traditional concept of free will and has significant implications for our understanding of moral responsibility and personal agency.

Overall, the debate over the nature of free will is complex and multifaceted, with different theories offering a range of perspectives on this fundamental aspect of human existence.

The implications of the idea that free will does or does not exist

The idea of free will has profound implications for how we view ourselves and our place in the world. If free will does exist, then we are responsible for our choices and actions, and we have the power to shape our own lives and destinies. This gives us a sense of agency and control over our lives, and allows us to take credit for our successes and learn from our mistakes.

On the other hand, if free will does not exist, then we are essentially automatons, controlled by the forces of nature and the laws of physics. Our actions and decisions are predetermined, and we have no real say in the course of our lives. This can be a deeply unsettling and disempowering idea, as it suggests that we are not really in control of our own lives.

The question of free will also has implications for morality and ethics. If we have free will, then we are responsible for our actions, and we can be held accountable for the harm we cause to others. If we do not have free will, then it becomes more difficult to assign blame or responsibility for actions, as they are ultimately determined by factors beyond our control.

Ultimately, the question of free will is deeply tied to our sense of identity and our understanding of what it means to be human. Whether we have free will or not, the concept has significant implications for how we view ourselves and our place in the world.

Chapter 13: The Meaning of Life

The search for the meaning of life is a fundamental human pursuit that has intrigued philosophers, theologians, scientists, and everyday people for millennia. What is the purpose of our existence? What is the ultimate goal of human life? These are questions that have been pondered by many, yet remain largely unanswered.

Throughout history, people have turned to religion, philosophy, and science to try to understand the meaning of life. Religion has provided answers based on faith and belief in a higher power, philosophy has explored the nature of existence and morality, and science has sought to understand the universe and our place in it.

However, despite the various attempts to provide an answer, the search for the meaning of life remains elusive. In this chapter, we will explore the different perspectives on the meaning of life and examine how they have evolved over time. We will delve into the philosophical, religious, and scientific approaches to the question and examine the challenges and limitations of each.

Ultimately, the meaning of life is a subjective concept that varies from person to person. While there may not be a single answer to the question, the exploration of different perspectives can provide insights into what gives our lives purpose and meaning.

The different theories of the meaning of life

The question of the meaning of life has puzzled humans for centuries, and it continues to be a subject of debate among philosophers, theologians, and scientists. There are many different theories about what the meaning of life is or should be.

One theory is that the meaning of life is to seek pleasure and avoid pain. This view has been advocated by various schools of philosophy, including hedonism and utilitarianism. According to this view, the goal of life is to maximize happiness and minimize suffering. However, critics of this view argue that pleasure-seeking can lead to a shallow and self-centered life, and that there are more important values than pleasure.

Another theory is that the meaning of life is to fulfill a particular purpose or destiny. This view has been advocated by various religions and spiritual traditions. According to this view, each person has a unique role to play in the world, and the goal of life is to discover and fulfill that role. However, critics of this view argue that it can lead to a narrow and dogmatic perspective, and that not everyone has a specific purpose in life.

A third theory is that the meaning of life is to create meaning. According to this view, there is no inherent meaning in life, but rather, it is up to each individual to create their own meaning through their experiences, relationships, and contributions to the world. This view has been advocated by existentialist philosophers and some psychologists. However, critics of this view argue that it can lead to a sense of nihilism and despair, and that there must be some objective meaning to life.

Ultimately, the question of the meaning of life may be unanswerable, or at least, there may not be a single answer that applies to everyone. Different people may find meaning in different things, such as love, art, knowledge, or spirituality. The search for meaning may be a lifelong journey, and one that is ultimately personal and subjective.

The implications of the idea that life has no inherent meaning

The idea that life has no inherent meaning can be both liberating and daunting. On one hand, it frees individuals from the burden of searching for an ultimate purpose or direction in life. One can create their own purpose and find fulfillment in their own personal goals and values. This can lead to a sense of personal autonomy and empowerment.

On the other hand, the idea that life has no inherent meaning can also lead to a sense of nihilism or despair. It may cause one to question the purpose of their existence and the value of their actions. Some may feel lost or directionless without a clear path to follow or a higher purpose to fulfill.

The implications of this idea can also extend beyond the individual level to society as a whole. It can challenge traditional religious and cultural beliefs that emphasize a predetermined purpose or destiny. It can also challenge societal structures and institutions that are based on certain assumptions about the meaning and purpose of life.

However, it is important to note that the absence of inherent meaning does not necessarily mean that life is meaningless. It simply means that individuals and society must create their own meaning and purpose. This can be a daunting task, but it can also be a source of inspiration and motivation to live a fulfilling and meaningful life.

The role of individual perspective in creating meaning

One of the most interesting aspects of the search for the meaning of life is the role that individual perspective plays in creating meaning. While there may not be a universally agreed-upon objective meaning of life, people have the capacity to find meaning in their own lives based on their own unique experiences and perspectives.

In fact, some philosophers and psychologists argue that the search for meaning is inherently subjective and that the meaning of life is something that individuals must create for themselves. From this perspective, the meaning of life is not something that can be discovered or uncovered, but rather something that is constantly being created through the choices and actions of individuals.

One way that individuals can create meaning in their lives is through their relationships with others. For many people, the bonds they form with family, friends, and romantic partners provide a sense of purpose and direction in life. Others find meaning through their careers or hobbies, pursuing interests and passions that give their lives a sense of fulfillment and purpose.

Ultimately, the meaning of life is something that each person must define for themselves. While there may be some broad philosophical or spiritual frameworks that can help guide this process, the search for meaning is a deeply personal and individual journey. By exploring their own values, interests, and relationships, individuals can create a sense of purpose and meaning that resonates with their own unique perspective on the world.

Chapter 14: The Future of Reality

The study of reality has been a central focus of human inquiry for centuries, and our understanding of reality has evolved significantly over time. In this chapter, we will explore what the future of reality might hold, given the current state of our knowledge and the various trends and developments that are shaping our world.

The first trend we should consider is the rapid pace of technological change. Advances in fields such as artificial intelligence, robotics, and virtual reality are transforming the way we interact with the world around us, blurring the line between the physical and the digital. As these technologies become more sophisticated and widespread, it is likely that they will continue to reshape our understanding of what is real and what is not.

Another trend that is likely to shape the future of reality is the ongoing debate over the nature of consciousness. As we continue to study the brain and learn more about how it generates subjective experience, it is possible that we will gain a deeper understanding of the relationship between consciousness and the physical world. This could have profound implications for our understanding of reality, particularly if it turns out that consciousness plays a more active role in shaping reality than we currently believe.

One possibility that has been explored in science fiction and philosophy alike is the idea that we are living in a simulation. If this is true, then our understanding of reality would be fundamentally different from what we currently believe, and the implications of this realization would be staggering. It is impossible to say for certain whether or not we are living in a simulation, but as our technology advances and we continue to explore the limits of what is possible, this idea may become more plausible.

Another possibility is that we will discover entirely new realms of reality beyond our current understanding. The study of dark matter and dark energy, for example, suggests that there may be vast expanses of the universe that we are currently unable to perceive. Similarly, the discovery of new dimensions or parallel universes could fundamentally alter our understanding of reality.

Finally, it is worth considering the role that human perception plays in shaping our understanding of reality. As we continue to study the brain and the mind, it is becoming increasingly clear that our perceptions of the world are deeply influenced by our expectations, biases, and past experiences. As we become more aware of these influences, we may be able to develop new tools and techniques for better understanding and navigating the world around us.

In conclusion, the future of reality is likely to be shaped by a complex array of factors, including technological change, scientific discovery, and human perception. While we cannot predict with certainty what the future will hold, it is clear that our understanding of reality is constantly evolving, and that we will continue to be surprised and challenged by the mysteries of the universe for years to come.

The potential implications of future technology on our understanding of reality

The future of reality is intrinsically linked with technological advancements. As we continue to develop new technologies, we will undoubtedly gain a deeper understanding of the nature of reality. In particular, emerging technologies such as virtual and augmented reality, artificial intelligence, and brain-computer interfaces have the potential to revolutionize our understanding of what is real.

Virtual and augmented reality technologies are rapidly advancing, providing increasingly immersive experiences that can replicate and simulate real-world environments. As these technologies continue to evolve, we may begin to question the very nature of reality, blurring the lines between what is real and what is not.

Artificial intelligence also has the potential to impact our understanding of reality. As we continue to develop more sophisticated AI systems, we may begin to question whether these systems possess their own consciousness and sense of reality. This could lead to new insights into the nature of consciousness and the mind-body problem.

Brain-computer interfaces, which allow for direct communication between the brain and external devices, also have the potential to challenge our understanding of reality. As these interfaces become more advanced, we may be able to create entirely new sensory experiences or even directly manipulate our perceptions of reality.

Overall, the potential implications of future technology on our understanding of reality are vast and unpredictable. While these advancements may challenge our current understanding of what is real, they may also open up new avenues for exploration and understanding. It will be up to us to navigate these changes and determine how best to use these technologies to further our understanding of the world around us.

The different scenarios for the future of reality

As technology advances at an unprecedented rate, many experts have speculated on what the future of reality might look like. Here are some of the potential scenarios:

1. Virtual Reality: With the development of increasingly sophisticated virtual reality technology, it is possible that people may spend more and more time in virtual worlds, which could blur the line between what is real and what is not. Some people may even prefer their virtual lives to their physical ones.

2. Augmented Reality: Augmented reality technology overlays digital information onto the physical world, creating a hybrid reality. This could change the way people experience the world around them, making it more immersive and interactive.

3. Artificial Intelligence: As artificial intelligence continues to advance, machines may become capable of complex decision-making and even develop consciousness. This could have profound implications for our understanding of reality, including the possibility that machines may create their own realities.

4. Transhumanism: Transhumanism is a movement that seeks to use technology to enhance human abilities and extend human life. This could lead to the creation of post-human beings with vastly different perspectives on reality.

5. Simulation Hypothesis: The simulation hypothesis suggests that reality is a computer simulation created by a more advanced civilization. As technology continues to advance, it is possible that we may one day create simulations indistinguishable from reality, leading to the possibility that our own reality could be a simulation.

These scenarios raise important questions about the nature of reality and the role of technology in shaping it. As we continue to push the boundaries of what is possible, it is likely that our understanding of reality will continue to evolve and transform.

The role of human consciousness in shaping the future of reality

The role of human consciousness in shaping the future of reality is a fascinating topic to consider. As we continue to develop new technologies and explore the limits of our understanding, it is becoming increasingly clear that our collective consciousness has a significant impact on the world around us.

One of the most significant ways in which human consciousness is shaping the future of reality is through the development of artificial intelligence. As we create machines that are capable of processing vast amounts of data and making complex decisions, we are essentially imbuing them with a form of consciousness. As these machines become more advanced, they will undoubtedly play an increasingly significant role in shaping the world around us.

Another way in which human consciousness is shaping the future of reality is through the development of new forms of communication and social interaction. With the advent of social media, we are now more connected than ever before, and our collective consciousness is becoming increasingly powerful. As we share information and ideas with one another, we are shaping the way in which we understand the world and our place within it.

At the same time, our individual consciousness is also playing an important role in shaping the future of reality. As we continue to learn and grow, we are constantly updating our understanding of the world around us. This ongoing process of personal growth and development is essential if we hope to create a better future for ourselves and for those around us.

Ultimately, the role of human consciousness in shaping the future of reality is a complex and multifaceted one. While our collective and individual consciousness will undoubtedly continue to play a significant role in shaping the world around us, it is important to remember that we are not the only factors at play. Other forces, such as technological progress and natural phenomena, will also have a significant impact on the future of reality. Nonetheless, by remaining mindful of our own consciousness and the way in which it shapes our understanding of the world, we can work towards creating a future that is more just, equitable, and sustainable for all.

Chapter 15: Conclusion

Throughout this book, we have explored some of the most fundamental and intriguing questions about the nature of reality. From the nature of time and space to the existence of free will and the meaning of life, these questions have captivated philosophers, scientists, and theologians for centuries.

We have seen that many of these questions remain open to debate, and that different theories and interpretations exist for each one. We have also seen that some of the most profound and surprising discoveries about the nature of reality have come from fields such as quantum mechanics, which challenge our everyday intuitions about how the world works.

Despite the many mysteries and uncertainties that remain, one thing is clear: our understanding of reality is constantly evolving. As we continue to explore the nature of the universe and our place within it, we will undoubtedly uncover new insights and ideas that challenge our existing beliefs and assumptions.

In this final chapter, we will reflect on what we have learned and consider some of the implications of these ideas for our lives and the future of humanity. We will also explore some of the exciting possibilities that lie ahead as we continue to push the boundaries of our understanding of reality.

Recap of the book's main points

Throughout this book, we have explored a variety of fascinating and thought-provoking ideas about the nature of reality. We have delved into topics such as the illusion of space, the role of consciousness in shaping our experience of reality, the concept of the multiverse, the possibility that our reality is a simulation, the nature of matter, and the mysteries of quantum mechanics. We have also examined the mind-body problem, the question of free will, and the meaning of life.

One of the main themes that has emerged throughout our exploration is the idea that our perception of reality is not necessarily a reflection of an objective reality "out there," but rather a construction of our minds. This idea challenges our conventional understanding of reality and raises important questions about the relationship between consciousness and the external world.

Another recurring theme is the notion that reality may be much more complex and multi-dimensional than we currently understand. We have explored the possibility of multiple realities, each with its own unique properties, and the potential implications of this idea for our understanding of the world around us.

Finally, we have examined the ways in which technology and human consciousness may continue to shape the future of reality. From the development of artificial intelligence and virtual reality to the ways in which we understand and interpret our experiences, the future of reality is a constantly evolving and endlessly fascinating subject.

Overall, this book has aimed to challenge our assumptions and encourage us to think deeply about the nature of reality. While we may never have all the answers, exploring these ideas can help us expand our understanding of the world and our place in it.

Final thoughts on the nature of reality and its implications

The study of reality and its various aspects is a never-ending pursuit. Throughout this book, we have explored various theories and ideas about the nature of reality, including the role of consciousness, the multiverse, the simulation hypothesis, the nature of matter, quantum mechanics, the mind-body problem, free will, and the meaning of life.

We have seen that reality is far more complex and nuanced than what we perceive with our senses, and that our understanding of it is constantly evolving. The debate over whether reality is objective or subjective, whether it exists independent of us or is a construct of our minds, and whether there are multiple realities or just one continues to shape our worldview and our understanding of the world.

Perhaps the most important takeaway from this book is that our perception of reality is deeply influenced by our individual perspectives and experiences. What may be true for one person may not be true for another. This highlights the importance of being open-minded, curious, and constantly questioning our assumptions and beliefs.

As we continue to explore the nature of reality and its implications, it is important to remember that our understanding of it is always limited by our current level of knowledge and understanding. It is up to us to continue to push the boundaries of our understanding, to explore new ideas and theories, and to embrace the unknown.

In conclusion, the study of reality is a never-ending journey, but it is one that is worth embarking on. By delving deeper into the mysteries of the universe, we can gain a greater appreciation for the complexity and beauty of the world around us, and ultimately, a greater understanding of ourselves.

Call to action for readers to continue exploring the truth about reality.

As you have journeyed through this exploration of the nature of reality, it is my hope that you have gained new insights and perspectives on the world around you. However, the search for truth is a never-ending process, and there is always more to learn and discover.

I encourage you to continue exploring the nature of reality, to question your assumptions and beliefs, and to seek out new ideas and perspectives. Engage in conversations with others who have different viewpoints, read books and articles on the subject, and attend lectures and events that address these topics.

By continuing to explore the truth about reality, we can deepen our understanding of ourselves and the world, and perhaps even contribute to shaping the future of reality itself. So, I urge you to take action and continue on this journey of discovery.